~A BINGO BOOK~

Our Solar System Bingo Book

COMPLETE BINGO GAME IN A BOOK

Source: NASA

Written By Rebecca Stark
Educational Books 'n' Bingo

TITLE: Our Solar System Bingo
AUTHOR: Rebecca Stark

ISBN 978-0-87386-440-4

Educational Books 'n' Bingo

Printed in the U.S.A.

OUR SOLAR SYSTEM BINGO DIRECTIONS

INCLUDED:

List of Terms

Templates for Additional Terms and Clues

2 Clues per Term

30 Unique Bingo Cards

Markers

1. **Either cut apart the book or make copies of ALL the sheets. You might want to make an extra copy of the clue sheets to use for introduction and review. Keep the sheets in an envelope for easy reuse.**

2. Cut apart the call cards with terms and clues.

3. Pass out one bingo card per student. There are enough for a class of 30.

4. Pass out markers. You may cut apart the markers included in this book or use any other small items of your choice.

5. Decide whether or not you will require the entire card to be filled. Requiring the entire card to be filled provides a better review. However, if you have a short time to fill, you may prefer to have them do the just the border or some other format. Tell the class before you begin what is required.

6. There are 50 terms. Read the list before you begin. If there are any terms that have not been covered in class, you may want to read to the students the term and clues before you begin.

7. There is a blank space in the middle of each card. You can instruct the students to use it as a free space or you can write in answers to cover terms not included. Of course, in this case you would create your own clues. (Templates provided.)

8. Shuffle the cards and place them in a pile. Two or three clues are provided for each term. If you plan to play the game with the same group more than once, you might want to choose a different clue for each game. If not, you may choose to use more than one clue.

9. Be sure to keep the cards you have used for the present game in a separate pile. When a student calls, "Bingo," he or she will have to verify that the correct answers are on his or her card AND that the markers were placed in response to the proper questions. Pull out the cards that are on the student's card keeping them in the order they were used in the game. Read each clue as it was given and ask the student to identify the correct answer from his or her card.

10. If the student has the correct answers on the card AND has shown that they were marked in response to the *correct questions,* then that student is the winner and the game is over. If the student does not have the correct answers on the card OR he or she marked the answers in response to *the wrong questions,* then the game continues until there is a proper winner.

11. If you want to play again, reshuffle the cards and begin again.

Have fun!

TERMS

asteroids	Mercury
astronaut(s)	meteoroid(s)
astronomer	Milky Way
astronomy	moon
atmosphere	NASA
axis	Neil Armstrong
big bang	Neptune
black hole	orbit(s)
comet/Comet	phases
constellation	planet(s)
astronomy	Pluto
crater	rings
axis	rocky
day	satellite(s)
Earth	seasons
eclipse	solar
galaxy / Galaxy	solar system
gas	space shuttle
gravity	star
John Glenn	sun
greenhouse effect	telescope
Jupiter	tide(s)
light year(s)	universe
lunar	Uranus
Mars	Venus

Additional Terms

Choose as many terms as you would like and write them in the squares.
Repeat each as desired. Cut out the squares and randomly
distribute them to the class.
Instruct the students to place the square on the center space of their card.

© **Barbara M. Peller**

Clues for
Additional Terms

Write two or three clues for each new term.

_____ 1. 2. 3.	_____ 1. 2. 3.
_____ 1. 2. 3.	_____ 1. 2. 3.
_____ 1. 2. 3.	_____ 1. 2. 3.

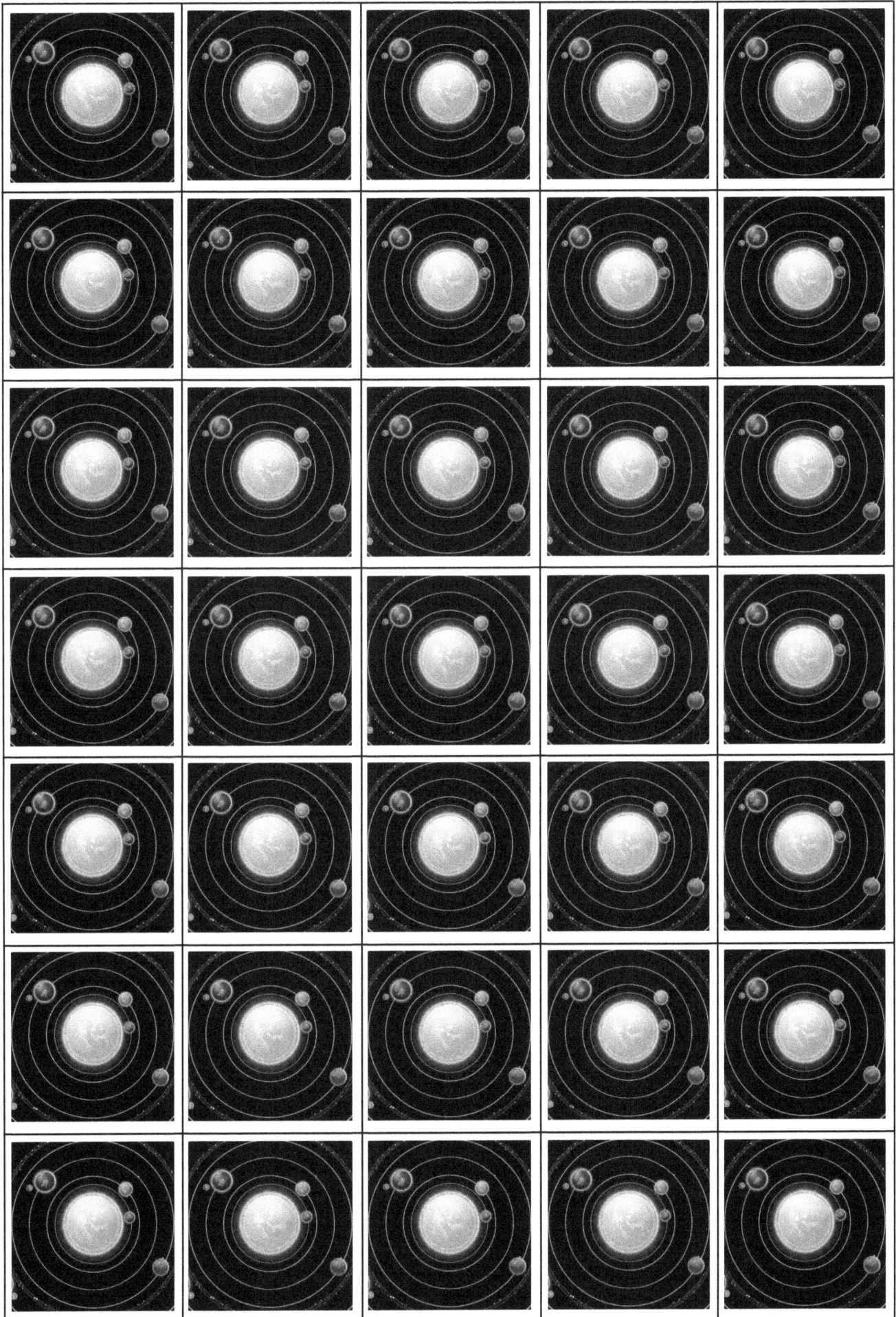

asteroids 1. These small celestial bodies are composed of rock and metal and orbit the sun. 2. These small metallic and rocky bodies orbit the sun mostly between the orbits of Mars and Jupiter.	**astronaut(s)** 1. A crew member in a spacecraft is an ___. 2. John Glenn and Neil Armstrong are famous ones.
astronomer 1. An ___ is a scientist who specializes in astronomy. 2. A scientist who studies celestial objects is called an ___.	**astronomy** 1. The study of objects and matters outside of Earth's atmosphere is called ___. 2. The branch of science that deals with stars, planets and other celestial objects is called ___.
atmosphere 1. The mass of air surrounding Earth is called its ___. 2. The mass of gas that surrounds a planet is that planet's ___.	**axis** 1. Earth rotates on its ___. 2. Earth's rotation on its ___ causes day and night.
big bang 1. This refers to the theory that a cosmic explosion marked the origin of the universe. 2. This theory hypothesizes that the universe began with an extremely powerful explosion.	**black hole** 1. A ___ results from the collapse of a star. 2. Things are pulled into a ___ because of gravity. Its mass is so great that nothing, including light, can escape.
comet/Comet 1. When close to the sun, it warms up and its parts can be seen. Haley's ___ is seen from Earth every 76 years. 2. Its parts include a nucleus, a coma, a hydrogen envelope, a dust tail, and an ion tail.	**constellation** 1. In everyday usage, we call a group of stars that form a pattern in the sky a ___. 2. The Big Dipper is one.

core 1. The central part of a planet is called its ___. 2. Earth and other terrestrial planets have a central metallic ___, composed mostly of iron.	**crater** 1. The bowl-shaped depression at the top of a volcano is a volcanic ___. 2. An impact ___ is the result of an impact of a meteoroid or other projectile on a planet's surface.
crust 1. The outermost solid layer of a planet or moon is called its ___. 2. Earth's ___ comprises the continents and ocean basins.	**day** 1. The time it takes a celestial body to rotate once on its axis is a ___. 2. Earth's ___ is 24 hours.
Earth 1. ___ is the third planet from the sun. 2. This is the only planet in our solar system where life is known to exist.	**eclipse** 1. A lunar ___ occurs when Earth comes between the sun and the moon, wholly or partially covering the moon. 2. A solar ___ occurs when the moon passes between the sun and Earth, wholly or partially covering the sun.
galaxy / Galaxy 1. Our sun and its solar system are part of the Milky Way ___. 2. Although the Andromeda ___ is closest to the Milky Way, it is about 2.4 million light years away.	**gas** 1. The Jovian planets are called ___ giants because they are composed mostly of hydrogen, helium, methane and ammonia. 2. There are two kinds of planets: rocky, or terrestrial, planets and Jovian, or ___, planets.
gravity 1. The force of attraction between all masses in the universe is called ___. 2. The planets revolve around the sun because of ___.	**John Glenn** 1. He was the first American to orbit Earth. 2. This astronaut orbited Earth aboard *Friendship 7* in 1962.

greenhouse effect 1. Gases in the atmosphere trap energy and prevent it from escaping back into space; this is the ___effect. 2. Carbon dioxide, nitrous oxide, methane and other gases act like a blanket that surrounds the planet. This is known as the ___.	**Jupiter** 1. It is the largest planet in our solar system. It is named for the ruler of the ancient Roman gods. 2. This planet's great red spot is actually a hurricane-like storm.
light year(s) 1. This unit of measurement is the distance light can travel in one year at a speed of 186,282 miles per second. 2. Our sun is 4.3 ___ from *Proxima Centauri,* the closest star.	**lunar** 1. ___ means "of or relating to the moon." 2. Complete this analogy: sun : solar :: moon : ___.
Mars 1. This is the fourth planet from the sun. 2. This planet is named after the ancient Roman god of war and is sometimes called the Red Planet.	**Mercury** 1. ___ is the closest planet to the sun. It orbits the sun once every 88 days. 2. ___ is the smallest planet in the solar system. It is named for the messenger god of the ancient Romans.
meteoroid(s) 1. Small bodies that travel through space are called ___. Most are pieces that have broken off from asteroids. 2. When it enters Earth's atmosphere, it is called a meteor. If it hits Earth, it is called a meteorite.	**Milky Way** 1. Our solar system is in this galaxy. 2. It is a spiral galaxy.
moon 1. A planet's natural satellite is called a ___. 2. Earth has one. Venus has none.	**NASA** 1. It is an acronym for National Aeronautics and Space Administration. 2. It was established in 1958. Its stated mission is "to pioneer the future in space exploration, scientific discovery and aeronautics research."
Our Solar System Bingo	

Neil Armstrong 1. He was the first human to set foot on the moon. Buzz Aldrin followed him. 2. This astronaut walked on the moon on July 20, 1969.	**Neptune** 1. It is the eighth planet from the sun. 2. This gas giant was named after the Roman god of the sea.
orbit(s) 1. The path of a heavenly body around another body is an ___. 2. The planets ___ around the sun; the moon ___ around Earth.	**phases** 1. The moon's appearance changes in stages from not visible to fully illuminated and back again. These stages are called ___. 2. The moon's 8 ___ are new moon, waxing crescent, 1st quarter, waxing gibbous, full moon, waning gibbous, last quarter & waning crescent.
planet(s) 1. A ___ is a body that orbits the sun and is large enough for its own gravity to make it round. 2. There are 8 ___ in our solar system. At one time there were said to be 9.	**Pluto** 1. ___ used to be considered the ninth planet in our solar system, but is now said to be too small. 2. ___ is considered a dwarf planet along with Eris and Ceres.
rings 1. The most spectacular planetary ___ are those around Saturn. 2. It is now known that Saturn is not the only one with ___ of cosmic dust orbiting around it. The other gas giants also have them.	**rocky** 1. The ___ inner planets are also called terrestrial planets. 2. The four ___ inner planets are Mercury, Mars, Venus and Earth.
satellite(s) 1. A ___ is an object that orbits a celestial body. 2. Moons are natural ___. Artificial ___ are manufactured objects that orbit Earth or another body in space. Our Solar System Bingo	**seasons** 1. The ___ are caused by the tilt of Earth's axis. 2. Because Earth's axis is tilted, different parts of the planet are oriented towards the sun at different times of the year, resulting in four ___. © Barbara M. Peller

solar 1. ___ means "of or relating to the sun." 2. Complete this analogy: moon : lunar :: sun : ___.	**solar system** 1. The ___ includes the sun, the planets, and the other celestial bodies that orbit the sun. 2. There are eight planets in our ___.
space shuttle 1. The ___ is a reusable vehicle that transports people and cargo between Earth and space. 2. The ___ *Challenger* had completed nine missions before it tragically broke apart during the launch of its tenth.	**star** 1. Our sun is a medium-sized yellow ___. 2. A ___ is a huge ball of gases held together by gravity. It radiates energy, some of which is released as visible light, making it glow.
sun 1. The ___ is at the center of our solar system. 2. The ___ is a medium-sized yellow star.	**telescope** 1. It is an optical instrument used for viewing distant objects. 2. A ___ consists of two or more lenses or mirrors.
tide(s) 1. The alternate rising and falling of the surface of the ocean are called ___. 2. There are two each day: high ___ and low ___.	**universe** 1. The ___ comprises all the space and matter in existence. 2. Complete this analogy: solar system : galaxy :: galaxy : ___.
Uranus 1. ___ is the seventh planet from the sun. 2. Like Neptune, ___ is called an ice giant, which is a sub-class of the gas giants.	**Venus** 1. ___ is the second planet from the sun. 2. This planet is named after the ancient Roman goddess of love.

Our Solar System Bingo

gas	avalanche	Jupiter	universe	solar system
black hole	gravity	sun	Milky Way	Mercury
satellite(s)	orbit(s)		greenhouse effect	moon
star	atmosphere	galaxy (Galaxy)	tide(s)	light year(s)
lunar	telescope	constellation	asteroids	John Glenn

Our Solar System Bingo

Our Solar System Bingo

universe	solar	Mars	Neptune	lunar
light year(s)	Earth	big bang	atmosphere	rocky
planet(s)	telescope		core	galaxy (Galaxy)
Milky Way	Pluto	orbit(s)	Venus	Mercury
John Glenn	sun	constellation	black hole	asteroids

Our Solar System Bingo

universe	galaxy (Galaxy)	Milky Way	tide(s)	satellite(s)
telescope	avalanche	astronomer	gravity	NASA
atmosphere	sun		rings	astronomy
orbit(s)	planet(s)	lunar	Earth	Mars
asteroids	constellation	black hole	Venus	Jupiter

Our Solar System Bingo

satellite(s)	tides	Milky Way	galaxy (galaxy)	universe
Earth	orbit	sooner	phase(s)	
shuttle	Venus		constellation	asteroids

Our Solar System Bingo

orbit(s)	rings	Jupiter	constellation	lunar
meteoroid(s)	Earth	gravity	Neptune	satellite(s)
greenhouse effect	big bang		solar system	tide(s)
galaxy (Galaxy)	eclipse	sun	black hole	astronomer
asteroids	John Glenn	Neil Armstrong	day	moon

© Barbara M. Peller

Our Solar System Bingo

John Glenn	solar system	atmosphere	big bang	constellation
meteoroid(s)	galaxy (Galaxy)	astronomer	orbit(s)	crust
solar	moon		avalanche	Jupiter
Mercury	rings	gas	Venus	day
Milky Way	black hole	phases	core	greenhouse effect

Our Solar System Bingo

astronomy	rings	Mars	solar	moon
tide(s)	atmosphere	day	gravity	satellite(s)
Neptune	astronomer		big bang	core
black hole	lunar	Venus	Neil Armstrong	greenhouse effect
light year(s)	galaxy (Galaxy)	gas	phases	Jupiter

Our Solar System Bingo

gas	rings	seasons	crust	Milky Way
light year(s)	Jupiter	telescope	avalanche	satellite(s)
Mars	tide(s)		core	crater
orbit(s)	Earth	meteoroid(s)	universe	planet(s)
constellation	black hole	Venus	Neil Armstrong	astronomer

Our Solar System Bingo

greenhouse effect	rings	axis	tide(s)	crater
meteoroid(s)	solar	Neptune	Jupiter	solar system
satellite(s)	rocky		moon	big bang
asteroids	orbit(s)	universe	day	Earth
sun	black hole	Neil Armstrong	atmosphere	light year(s)

Our Solar System Bingo

core	Milky Way	telescope	satellite(s)	moon
day	solar	greenhouse effect	atmosphere	Jupiter
NASA	gas		avalanche	axis
crater	John Glenn	lunar	crust	seasons
Earth	Venus	astronomy	universe	solar system

Our Solar System Bingo

star	universe	big bang	Neptune	phases
moon	crater	gravity	avalanche	Jupiter
rings	rocky		tide(s)	planet(s)
lunar	Mercury	day	Venus	NASA
comet/Comet	John Glenn	Mars	light year(s)	greenhouse effect

Our Solar System Bingo

astronomer	rocky	atmosphere	day	light year(s)
axis	NASA	crust	core	gravity
meteoroid(s)	solar		Mars	telescope
comet/Comet	satellite(s)	Venus	black hole	universe
astronomy	constellation	gas	Neil Armstrong	Milky Way

Our Solar System Bingo

Milky Way	Earth	NASA	tide(s)	core
telescope	sun	solar	Neil Armstrong	meteoroid(s)
gas	seasons		moon	Neptune
constellation	solar system	Jupiter	universe	avalanche
rocky	axis	rings	astronomer	crater

Our Solar System Bingo

comet/Comet	solar system	astronomy	NASA	moon
solar	axis	rings	core	planet(s)
tide(s)	big bang		telescope	seasons
greenhouse effect	Venus	crater	rocky	universe
black hole	Mercury	Neil Armstrong	gas	crust

Our Solar System Bingo

constellation	solar	atmosphere	core	comet/Comet
crater	gas	NASA	avalanche	planet(s)
day	tide(s)		Mars	big bang
Mercury	Venus	rings	astronomer	astronomy
black hole	Neptune	rocky	light year(s)	greenhouse effect

Our Solar System Bingo

crust	core	atmosphere	Milky Way	Jupiter
astronomy	phases	gravity	solar	day
moon	gas		satellite(s)	tide(s)
black hole	NASA	axis	Venus	comet/Comet
light year(s)	Earth	Neil Armstrong	Mars	telescope

Our Solar System
Bingo

Jupiter	Milky Way	atmosphere	core	
	comet			Black hole
telescope		del Amazonas	sun	

Our Solar System Bingo

big bang	Uranus	axis	phases	Pluto
Neptune	rocky	seasons	meteoroid(s)	star
comet/Comet	solar system		moon	telescope
orbit(s)	Earth	black hole	crust	universe
day	NASA	Neil Armstrong	crater	planet(s)

Our Solar System Bingo

comet/Comet	space shuttle	eclipse	NASA	black hole
crust	day	Venus	tide(s)	seasons
core	star		Uranus	axis
John Glenn	light year(s)	greenhouse effect	atmosphere	planet(s)
lunar	astronomer	Milky Way	universe	solar system

Our Solar System Bingo

Jupiter	rings	crater	day	Neptune
John Glenn	comet/Comet	atmosphere	moon	astronomer
core	planet(s)		eclipse	phases
rocky	gravity	Venus	star	Mars
Uranus	NASA	lunar	space shuttle	astronomy

Our Solar System Bingo: Card No. 18

Our Solar System Bingo

moon	astronomy	NASA	axis	rocky
crust	constellation	phases	Milky Way	star
space shuttle	tide(s)		avalanche	atmosphere
Mars	Uranus	lunar	Earth	eclipse
satellite(s)	Pluto	light year(s)	greenhouse effect	Neil Armstrong

Our Solar System Bingo

rocky	space shuttle	star	NASA	avalanche
big bang	telescope	meteoroid(s)	lunar	Neptune
solar system	seasons		orbit(s)	eclipse
John Glenn	greenhouse effect	asteroids	Earth	Uranus
galaxy (Galaxy)	sun	Pluto	universe	gravity

Our Solar System Bingo

astronomy	John Glenn	meteoroid(s)	NASA	Mercury
solar system	eclipse	crater	axis	gas
planet(s)	light year(s)		space shuttle	atmosphere
lunar	Milky Way	Uranus	crust	greenhouse effect
orbit(s)	Pluto	Neil Armstrong	comet/Comet	Earth

Our Solar System Bingo

satellite(s)	Mars	eclipse	solar	comet/Comet
Neptune	star	Jupiter	axis	avalanche
crater	tide(s)		gas	seasons
Uranus	John Glenn	Earth	gravity	constellation
Pluto	astronomer	space shuttle	planet(s)	meteoroid(s)

Our Solar System Bingo

big bang	space shuttle	Milky Way	solar	Neil Armstrong
astronomy	rocky	light year(s)	crust	gravity
Mars	comet/Comet		asteroids	gas
planet(s)	sun	Uranus	astronomer	Earth
Mercury	greenhouse effect	Pluto	lunar	eclipse

Our Solar System Bingo

Neil Armstrong	solar	Milky Way	space shuttle	big bang
		asteroid		
Earth		Uranus	sun	planet(s)
eclipse	Jupiter	Pluto	greenhouse effect	Mercury

Our Solar System Bingo

big bang	rocky	constellation	space shuttle	axis
moon	Neil Armstrong	meteoroid(s)	Neptune	gas
seasons	phases		comet/Comet	planet(s)
Mercury	asteroids	Uranus	astronomer	solar system
galaxy (Galaxy)	orbit(s)	Pluto	star	sun

Our Solar System Bingo

orbit(s)	meteoroid(s)	space shuttle	atmosphere	eclipse
gravity	Mercury	crust	big bang	avalanche
solar system	axis		asteroids	Uranus
phases	John Glenn	sun	Pluto	star
Neil Armstrong	constellation	crater	day	galaxy (Galaxy)

Our Solar System Bingo: Card No. 25

Our Solar System Bingo

eclipse	space shuttle	asteroids	Neptune	phases
Mars	tide(s)	axis	rocky	big bang
Mercury	lunar		star	orbit(s)
comet/Comet	solar	John Glenn	Pluto	Uranus
seasons	day	atmosphere	sun	galaxy (Galaxy)

Our Solar System Bingo

asteroids	crater	space shuttle	rocky	telescope
Mercury	Mars	crust	Uranus	avalanche
Venus	sun		Pluto	orbit(s)
phases	astronomy	galaxy (Galaxy)	meteoroid(s)	gravity
comet/Comet	star	eclipse	satellite(s)	seasons

Our Solar System
Bingo

Our Solar System Bingo

moon	rings	universe	space shuttle	crater
telescope	eclipse	asteroids	lunar	star
sun	planet(s)		phases	Neptune
seasons	satellite(s)	light year(s)	Pluto	Uranus
solar	core	comet/Comet	galaxy (Galaxy)	Mercury

Our Solar System Bingo

eclipse	rings	phases	crust	core
Mercury	lunar	meteoroid(s)	seasons	satellite(s)
solar system	space shuttle		avalanche	asteroids
telescope	John Glenn	Jupiter	Pluto	Uranus
big bang	axis	galaxy (Galaxy)	astronomy	sun

Our Solar System Bingo

constellation	space shuttle	Neptune	core	Uranus
gravity	phases	Mars	star	avalanche
galaxy (Galaxy)	sun		seasons	meteoroid(s)
Mercury	astronomer	eclipse	Pluto	asteroids
John Glenn	Milky Way	astronomy	rings	Jupiter

www.ingramcontent.com/pod-product-compliance
Lightning Source LLC
Chambersburg PA
CBHW051420200326
41520CB00023B/7302